1000 Dormers

Jo Cryder

4880 Lower Valley Road, Atglen, Pennsylvania 19310

Copyright © 2007 by Jo Cryder
Library of Congress Control Number: 2007928130

All rights reserved. No part of this work may be reproduced or used in any form or by any means—graphic, electronic, or mechanical, including photocopying or information storage and retrieval systems—without written permission from the publisher.
The scanning, uploading and distribution of this book or any part thereof via the Internet or via any other means without the permission of the publisher is illegal and punishable by law. Please purchase only authorized editions and do not participate in or encourage the electronic piracy of copyrighted materials.
"Schiffer," "Schiffer Publishing Ltd. & Design," and the "Design of pen and ink well" are registered trademarks of Schiffer Publishing Ltd.

Designed by "Sue"
Type set in Bernhard Modern BT/Souvenir Lt BT

ISBN: 978-0-7643-2710-0
Printed in China

Published by Schiffer Publishing Ltd.
4880 Lower Valley Road
Atglen, PA 19310
Phone: (610) 593-1777; Fax: (610) 593-2002
E-mail: Info@schifferbooks.com

For the largest selection of fine reference books on this and related subjects, please visit our web site at **www.schiffer-books.com**
We are always looking for people to write books on new and related subjects. If you have an idea for a book please contact us at the above address.

This book may be purchased from the publisher.
Include $3.95 for shipping.
Please try your bookstore first.
You may write for a free catalog.

In Europe, Schiffer books are distributed by
Bushwood Books
6 Marksbury Ave.
Kew Gardens
Surrey TW9 4JF England
Phone: 44 (0) 20 8392-8585; Fax: 44 (0) 20 8392-9876
E-mail: info@bushwoodbooks.co.uk
Website: www.bushwoodbooks.co.uk
Free postage in the U.K., Europe; air mail at cost.

Contents

Introduction 4
Sketches of Dormer Designs 23
Photographs of Dormers 24
Section 1. Gable Dormers 24
    Section 1-a. One Window Gable Dormer 24
    Section 1-b. Two Window Gable Dormer 41
    Section 1-c. Three Window Gable Dormer 46
    Section 1-d. Four Window Gable Dormer 48
    Section 1-e. Enclosed Eave Gable Dormer 49
    Section 1-f. Wrap-around Enclosed Eave Gable Dormer 59
    Section 1-g. "Pediment" or "Triangle" Gable Dormer 64
    Section 1-h. Double Gable Dormer 75
Section 2. Hipped Dormers 78
Section 3. Dormers with Curves 98
    Section 3-a. Arched Dormer 98
    Section 3-b. Round or Oval Dormer 103
    Section 3-c. Eyebrow Dormer 104
Section 4. "Pediment" or "Triangle" Dormers 108
Section 5. "Flat", "Box" or "Shed" Dormers 123
Section 6. Turret Dormers 139
Section 7. Deck Dormers 144
Section 8. Inset Dormers 155
Section 9. "Through-the-Cornice" or "Wall" Dormers 157
Section 10. Special Mention Dormers 184

Introduction

A dormer is a roof projection usually containing a window or vent. Reasons to add dormers during construction or remodeling are:

1. Dormers add space. They append windows and vents that bring in light and ventilation.
2. A dormer adds character and dimension to the inside and outside of a structure.
3. A dormer is an architectural alternative to adding skylights and roof vents.
4. A dormer can be used to provide an outside access.
5. Dormers match or augment other features on the structure, such as turrets, cupolas, belvederes, and the rooflines.

The Fulton Mansion. Rockport, Texas. Read more about the Fulton Mansion at the following website: http://www.rockport-fulton.org/frames/mansion.htm

Originally, dormers were added to enhance interior attic space that would otherwise be good for little but storage. As architecture developed, dormers started being used to augment most any space in a structure. It is important that the design and location of a dormer be well thought out.

Mansions and Their Dormers

A myriad of designs and sizes of dormers exists to choose from. When proportionately added to a structure they can transform an unadorned roof into a crown. To introduce you to dormers, here are a few grand old mansions that have dormers along with other "jewels" in their crown. I have identified most of these mansions, but will not identify the structures where the rest of the dormers are located. We are going to concentrate on the dormers and their designs.

Rockport, Texas

The Ashton Villa. Galveston, Texas. To see more of the historic homes in Galveston, go to: http://www.youtube.com/watch?v=1VnTPaUwREc

The Moody Mansion. Galveston, Texas

Galveston, Texas

Bishop's Palace. Galveston, Texas

More Old-line Dormers

The following pictures are close-ups of old-line dormers used on homes and buildings built at an earlier time. The dormers are embellished with ornamentation and enhance each structure they adorn.

Galveston, Texas

Galveston, Texas

Key West, Florida

Key West, Florida

Key West, Florida

Georgetown, Kentucky

Evansville, Indiana

Evansville, Indiana

Evansville, Indiana

Evansville, Indiana

Terre Haute, Indiana

Bloomington, Illinois

Bloomington, Illinois

Terre Haute, Indiana

Newer Houses with Dormers

Dormers have continued to develop in shape, size, purpose, and location. As architecture has evolved, the use and shapes of dormers has expanded. Here are examples of some more recently constructed homes with dormers.

Brownsville, Texas

Corpus Christi, Texas

Brownsville, Texas

Galveston Island, Texas

13

Galveston Island, Texas

Galveston Island, Texas

Florida Keys

Key West, Florida

Savannah, Georgia

Flat Rock, North Carolina

15

Pisgah, Kentucky

Dallas, Oregon

Janesville, Wisconsin

Variety of Dormers on a House

Generally speaking, it is more common for structures to have matching dormers. However, the following examples illustrate a variety of designs and sizes combined on the same structure. The results show how this can heighten the overall affect.

Lewisville, Texas

Gatesville, Texas

Crawford, Texas

Corpus Christi, Texas

Crawford, Texas

Aransas Pass, Texas

Palacios, Texas

Key West, Florida

Fallets Island, Texas

Key West, Florida

Galveston Island, Texas

Key West, Florida

Key West, Florida

Clinton, Tennessee

Charleston, South Carolina

Winchester, Kentucky

Folly Beach, South Carolina

Winchester, Kentucky

Princeton, Indiana

Danville, Illinois

Bloomington, Illinois

Bloomington, Illinois

Monroe, Wisconsin

Rockton, Illinois

Dallas, Oregon

Dormer Classification

I'm not an architect or designer, and I certainly don't profess to be an authority on dormers. What I'm offering here is a sampling of dormers that already exist. Locating them so that I could photograph them was one thing, but identifying and sorting them was quite another. During my research, I found that dormer designs are not always identified with the same name. If the designs have different names in different references, I've listed each name I found. In the cases where dormer designs I located were not included, I've given them names. Sketches are provided as a guide to the general shapes and categories into which the dormers have been sorted.

The roof of the dormer is the main defining element. However, dormers that are of the same design can look different from each other. As you'll see in the following pages, some of the features that can cause dormers of the same design to be different from each other are:

- Overall size of the dormer—height, width, depth
- Pitch of the building's roof
- Location of the dormer on the building's roof
- Pitch of the roof on the dormer
- Size of the roof and eave overhang
- Number, size, style, and shape of windows and/or vents, or the lack of either
- Colors and materials
- Decoration, style or finish on the roof, eaves, windows, and walls
- Purpose of the dormer

These criteria will be repeated in each section to remind you to watch for these distinctions as you compare the dormers. Please note that the photographs are accurate and untouched. Any comments are my own opinions.

Sketches of Dormer Designs

This page contains sketches of the various designs of dormers. Each sketch is identified with the section number where the pictures of that particular design can be found in this book.

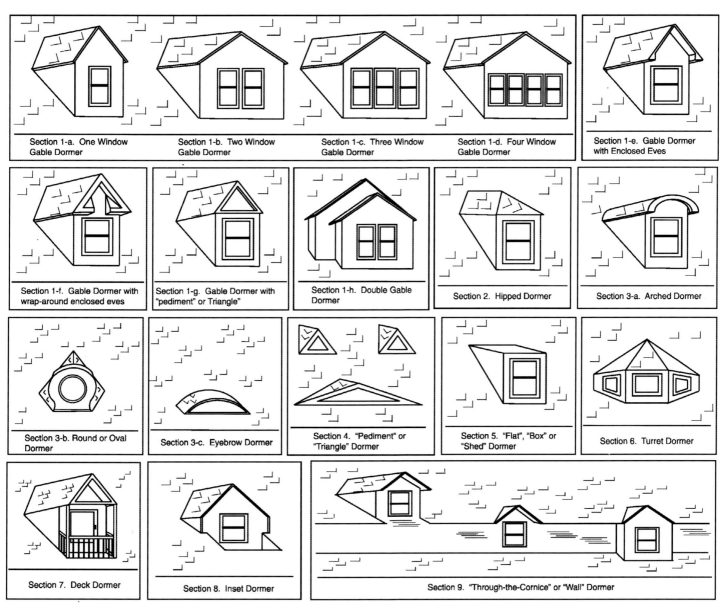

Photographs of Dormers

The dormers in Sections 1-a through 6 are sorted by design, which is often established by roof design. The dormers in Sections 7 through 10 are sorted by their special features. When a type of dormer is plentiful, I've sorted them by the number of windows they contain. The city and state where the pictures were taken are shown under each photograph simply for orientation. Within each section the pictures appear in the order in which they were taken.

Section 1. Gable Dormers

Gable dormers are identified by their roof lines built in an "A" shape. These dormers are popular, so there are many examples and varieties. Because of the large quantity of gable dormers, they are additionally sorted as follows: dormers that have one window, two windows, three windows, and four windows; enclosed eaves; wrap-around enclosed eaves; pediment; and double gable.

Section 1-a. One Window Gable Dormer

One window gable dormers are popular. They may be used solely to provide light and/or air circulation. Whatever the reason, they add dimension to the exterior and additional light, air, and space to the interior.

Lewisville, Texas

Lewisville, Texas. These dormers have a roof overhang on the front that provides beneficial shading from the sun and rain.

Crawford, Texas

Crawford, Texas. Notice how close the front dormer is to the edge of the roof.

Crawford, Texas

Gatesville, Texas

Zapata, Texas

Brownsville, Texas

Fallets Island, Texas

Brownsville, Texas

Fallets Island, Texas. This is a fairly large dormer with one small window. Later you will see similar sized dormers with two or more windows.

Fallets Island, Texas

Galveston, Texas

Florida Keys. Notice the location of the dormer at the peak of the roof, the added decoration and the Bahama or Bermuda shutter.

Florida Keys

Key West, Florida. This dormer is located right on the edge of the cornice. Also notice the colonial style shutters.

Florida Keys. Gable dormer.

Florida Keys. One vent gable dormer

Key West, Florida

Key West, Florida. This picture was taken while roof damage from the previous hurricane season was being repaired.

Savannah, Georgia

Savannah, Georgia

29

Savannah, Georgia

Savannah, Georgia. Gable dormer.

Savannah, Georgia

Charleston, South Carolina. These dormers are located at the peak of the building's roof. They have large casement windows.

Savannah, Georgia

Folly Beach, South Carolina

Folly Beach, South Carolina

Folly Beach, South Carolina

Flat Rock, North Carolina

Hendersonville, North Carolina. The interesting thing about these dormers is that the roofs are not the same material as the structure's roof.

Ashville, North Carolina

Clinton, Tennessee

Rarity Ridge near Oak Ridge, Tennessee

Winchester, Kentucky. This window has an awning and decorative flower box.

Winchester, Kentucky. Notice how the dormer has no eaves or overhang.

Winchester, Kentucky. This dormer has extended eaves and overhang plus a great window.

Lexington, Kentucky. Gable dormer. It's hard to determine if this dormer has a window with a permanent, custom built-shutter or if this is a vent.

Winchester, Kentucky

Lexington, Kentucky. The low pitch of this building's roof causes the dormer to be deeper.

Turkey Run State Park, Indiana. This dormer is tucked in under the eave of the second floor.

Bloomington, Illinois. The dormer has large decorative colonial shutters.

Bloomington, Illinois. Gable dormer.

Orfordville, Wisconsin

Beloit, Wisconsin

Shorewood (Milwaukee), Wisconsin

Shorewood (Milwaukee), Wisconsin

House on the Rock, Wisconsin. The window is recessed in this dormer.

Rockford, Illinois

Rockford, Illinois

Rockford, Illinois

Rockford, Illinois

Wisconsin Dells, Wisconsin

Wisconsin Dells, Wisconsin

Wisconsin Dells, Wisconsin

Wisconsin Dells, Wisconsin

Wisconsin Dells, Wisconsin

Dallas, Oregon

Section 1-b. Two Window Gable Dormer

The size of dormers with two windows is usually larger than those with one window. This increases the amount of usable space, light, and air circulation.

Florida Keys

Gatesville, Texas

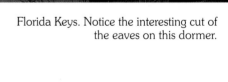

Florida Keys. Notice the interesting cut of the eaves on this dormer.

Florida Keys. Vent and two window gable dormer.

Ashville, North Carolina

Evansville, Indiana

Bloomington, Illinois

Frankfort, Kentucky. The pitch of the roof and the decoration on the front is appealing.

Frankfort, Kentucky

Bloomington, Illinois

Bloomington, Illinois

Bloomington, Illinois. Notice the symmetry of the windows on the dormer with the windows below. Behind the dormer at the peak of the building's roof is a belvedere.

Oak Park (Chicago), Illinois

Walworth, Wisconsin. Behind the dormer, at the peak of the building's roof, is a turreted belvedere.

House on the Rock, Wisconsin. This dormer has another dormer on top of it.

Wisconsin Dells, Wisconsin

Dallas, Oregon

Dallas, Oregon. Gable dormer. This dormer has no windows on the front and decoration has been added. Look closely and you will see that the windows are on the sides of the dormer.

Section 1-c. Three Window Gable Dormer

The size of dormers with three windows increases exponentially the amount of usable space, light, and air circulation.

Florida Keys

Key West, Florida

Charleston, South Carolina

Hendersonville, North Carolina

Winchester, Kentucky

Winchester, Kentucky

Versailles, Kentucky

Section 1-d. Four Window Gable Dormer
Dormers with four windows again have an increased amount of usable space, light, and air circulation.

Florida Keys

Winchester, Kentucky

Rockford, Illinois

Section 1-e. Gable Dormer with Enclosed Eaves

Enclosing the eaves changes the appearance of the dormer and gives it a clean-cut appearance. It also makes painting and maintenance of the eaves easier and can help reduce the incidence of hornet and birds' nests.

Lewisville, Texas

Lopeno, Texas

Mission, Texas

Mission, Texas. These dormers are lined up next to each other and positioned at one end of the roof.

Brownsville, Texas. Five matching dormers across the length of the roof.

Los Fresnos, Texas

Aransas Pass, Texas

Fallets Island, Texas. This is the only dormer I found with this roof line and feature in the front, plus windows on each side. See Section 10 for more unique dormers.

Brunswick, Georgia. These dormers appear to have vents instead of windows.

Galveston Island, Texas

Hilton Head, South Carolina

Hilton Head, South Carolina. A pair of working colonial shutters hides the window.

Folly Beach, South Carolina. A railing added in front of the dormer dresses it up.

Charleston, South Carolina

Hendersonville, North Carolina

Hendersonville, North Carolina

Ashville, North Carolina

Clinton, Tennessee

Clinton, Tennessee

Clinton, Tennessee

Winchester, Kentucky

Georgetown, Kentucky

Evansville, Indiana

Evansville, Indiana

55

Fort Branch, Indiana

Fort Branch, Indiana. This dormer has vents on either side of the window.

Bloomington, Illinois

Beloit, Wisconsin

Sharon, Wisconsin

Rockford, Illinois

Wisconsin Dells, Wisconsin

Wisconsin Dells, Wisconsin

Wisconsin Dells, Wisconsin

Dallas, Oregon

Section 1-f. Gable Dormer with Wrap-around Enclosed Eaves

Wrap-around eaves are an extension of enclosed eaves and provide one more way of finishing off the dormer. The enclosed eaves wrap around the front of the gable dormer and at times are covered with roofing material. Often they match other rooflines of the house.

Belton, Texas

Brownsville, Texas

Palacios, Texas. Notice the way the roof line of the dormer merges with the roof line of the building.

Galveston Island, Texas

Charleston, South Carolina

Clinton, Tennessee

Ashville, North Carolina

Clinton, Tennessee

Clinton, Tennessee

Lexington, Kentucky

Frankfort, Kentucky

Georgetown, Kentucky

Fort Branch, Indiana

Oak Park (Chicago), Illinois

Wisconsin Dells, Wisconsin

Section 1-g. Gable Dormer with Pediment or Triangle

These gable dormers have a pediment or triangle at the upper front of the face inside the "A". In some cases, the pediment is left open at the apex of the sloping cornice, or broken along the base. Some dormers shown here are extravagantly decorated.

Pearsall, Texas

Falcon, Texas

Corpus Christi, Texas. The pediment has been hipped along the bottom line and finished off with roofing material. Then, that bottom line is extended out in front. The eaves are extended as well. The shading on the front of the dormer and the lack of stain on the roofing material show how this protects the dormer underneath.

Galveston Island, Texas

Galveston, Texas

Galveston, Texas

Galveston, Texas

Key West, Florida

Charleston, South Carolina

Charleston, South Carolina

Charleston, South Carolina

Hendersonville, North Carolina

Flat Rock, North Carolina

Ashville, North Carolina

Ashville, North Carolina

Ashville, North Carolina

Winchester, Kentucky

Winchester, Kentucky

Winchester, Kentucky

Lexington, Kentucky

Lexington, Kentucky. This dormer pediment or triangle is open and has a column.

Georgetown, Kentucky

Evansville, Indiana

Terre Haute, Indiana. Notice the use of square columns and the pediment or triangle.

Evansville, Indiana

71

Bloomington, Illinois

Bloomington, Illinois

Bloomington, Illinois

Bloomington, Illinois

Bloomington, Illinois

Beloit, Wisconsin

Bloomington, Illinois

Beloit, Wisconsin

Lake Geneva, Wisconsin

Rockford, Illinois

Rockford, Illinois

Section 1-h. Double Gable Dormer

Double gable dormers are constructed with two or three dormers that are "stacked" one in front of the other.

South Padre Island, Texas

Clinton, Tennessee

Georgetown, Kentucky

Danville, Illinois

Dallas, Oregon

Dallas, Oregon

Dallas, Oregon

Dallas, Oregon

Dallas, Oregon

Section 2. Hipped Dormers

A hipped roof has a change in roof direction, where two planes meet at an angle to make a vertical ridge or fold. The roof slopes down to the eaves on three sides.

Because of the quantity of hipped dormers presented here, they are additionally sorted as follows: dormers that have one window, two windows, three windows, and four windows.

Brownsville, Texas

Fallets Island, Texas

Pearsall, Texas

Aransas Pass, Texas

Florida Keys

Florida Keys

Folly Beach, South Carolina

Charleston, South Carolina

Charleston, South Carolina

Charleston, South Carolina

Charleston, South Carolina

Charleston, South Carolina

Charleston, South Carolina

Charleston, South Carolina

Ashville, North Carolina

Ashville, North Carolina

Ashville, North Carolina

Ashville, North Carolina

Ashville, North Carolina

Ashville, North Carolina

Lexington, Kentucky

Evansville, Indiana

Fort Branch, Indiana

Bloomington, Illinois

Shorewood (Milwaukee), Wisconsin

Dallas, Oregon

South Padre Island, Texas

Corpus Christi, Texas

Galveston, Texas

Key West, Florida

Key West, Florida

Key West, Florida

Savannah, Georgia

Savannah, Georgia

Beaufort, South Carolina

Beaufort, South Carolina

Charleston, South Carolina

Charleston, South Carolina

Charleston, South Carolina

Ashville, North Carolina

Clinton, Tennessee

Rarity Ridge near Oak Ridge, Tennessee

Winchester, Kentucky

Winchester, Kentucky

Winchester, Kentucky

Winchester, Kentucky

Winchester, Kentucky

Winchester, Kentucky

Winchester, Kentucky

Versailles, Kentucky

Frankfort, Kentucky

Georgetown, Kentucky

Evansville, Indiana

Evansville, Indiana

Bloomington, Illinois

Oak Park (Chicago), Illinois

Oak Park (Chicago), Illinois

Beloit, Wisconsin

Rockford, Illinois

Dallas, Oregon

Palacios, Texas

Dallas, Oregon

Savannah, Georgia

Charleston, South Carolina

Key West, Florida

Charleston, South Carolina

Charleston, South Carolina

Folly Beach, South Carolina

Charleston, South Carolina

Winchester, Kentucky

Georgetown, Kentucky

Winchester, Kentucky

Evansville, Indiana

Evansville, Indiana

Bloomington, Illinois

Beloit, Wisconsin

Charleston, South Carolina

Winchester, Kentucky

Lexington, Kentucky

Section 3. Dormers with Curves

The dormers in this section all contain curved roofs or shapes. The appeal of the curves, strength of the arch, and enhanced water run-off make these dormers appealing. These dormer styles are often embellished with unique windows, ornamentation, or other details.

Section 3-a. Arched Dormers

Arched dormers are attention-getters. They can look good when used on structures with arched windows and add curves to structures that are otherwise all angles.

Lewisville, Texas

Rockport, Texas

Mission, Texas

Savannah, Georgia

Charleston, South Carolina

Hendersonville, North Carolina

Ashville, North Carolina

Clinton, Tennessee

Lexington, Kentucky. This dormer matches the next dormer in design, except that this one has a vent.

Clinton, Tennessee

Lexington, Kentucky. This dormer matches the previous dormer in design, except that this one has a window.

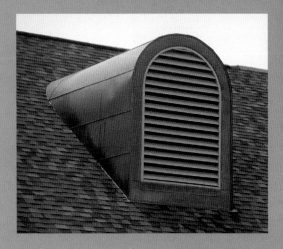

Lexington, Kentucky

Lexington, Kentucky

Lexington, Kentucky

Frankfort, Kentucky

Bloomington, Illinois. An arched dormer tucked in between two chimneys.

Bloomington, Illinois

Rockford, Illinois

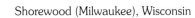

Shorewood (Milwaukee), Wisconsin

Rockford, Illinois

Section 3-b. Round or Oval Dormers

Round and oval dormers are unique. In my travels I found only a few examples to photograph.

Lewisville, Texas

Savannah, Georgia

Savannah, Georgia

Low country near Beaufort, South Carolina

Low country near Beaufort, South Carolina

Section 3-c. Eyebrow Dormers

Eyebrow dormers are very eye-catching and can lend a mysterious aura to a structure.

Beaufort, South Carolina

Galveston, Texas

Charleston, South Carolina

Ashville, North Carolina

Frankfort, Kentucky

Bloomington, Illinois

Lexington, Kentucky

Bloomington, Illinois

105

Shorewood (Milwaukee), Wisconsin

Rockford, Illinois

Rockford, Illinois

Rockford, Illinois

Rockton, Illinois

Rockton, Illinois. This shows the appearance of the inside of the dormer in the previous picture.

Dallas, Oregon

Section 4. "Pediment" or "Triangle" Dormer

The word pediment comes from the Latin "pedare," which means "to support." This dormer also gets its name from the triangle shape. Unlike the gable dormer with pediment or triangle shown earlier, these dormers are made with a pediment or triangle only—no other walls.

Lewisville, Texas

Pearsall, Texas

Gatesville, Texas

Gatesville, Texas. Notice the hipped roof on this dormer.

Mission, Texas

Corpus Christi, Texas

Galveston Island, Texas

Aransas Pass, Texas. Notice the hipped roof on this dormer.

Galveston Island, Texas

Galveston Island, Texas

Galveston, Texas

Galveston, Texas

Florida Keys

Galveston, Texas

Key West, Florida

Florida Keys

Key West, Florida

Key West, Florida

Key West, Florida

Key West, Florida

111

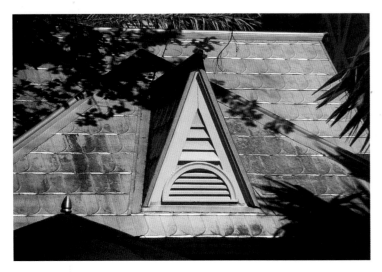

Key West, Florida

Florida Keys

Florida Keys

Florida Keys

Key West, Florida

Key West, Florida

Key West, Florida

Key West, Florida

Savannah, Georgia

Savannah, Georgia

Hilton Head, South Carolina

Beaufort, South Carolina

Hilton Head, South Carolina

Charleston, South Carolina

Hilton Head, South Carolina

Hendersonville, North Carolina

Flat Rock, North Carolina

Hendersonville, North Carolina

Clinton, Tennessee

Hendersonville, North Carolina

Clinton, Tennessee

Clinton, Tennessee

Winchester, Kentucky

Winchester, Kentucky

Corbin, Kentucky

Clinton, Tennessee

Corbin, Kentucky

117

Winchester, Kentucky

Winchester, Kentucky

Winchester, Kentucky

Winchester, Kentucky

Winchester, Kentucky

Winchester, Kentucky

Winchester, Kentucky

Winchester, Kentucky

Winchester, Kentucky

Winchester, Kentucky

Winchester, Kentucky

Frankfort, Kentucky

Frankfort, Kentucky

Lexington, Kentucky

Evansville, Indiana

Bloomington, Illinois

Bloomington, Illinois

Bloomington, Illinois

Lake Geneva, Wisconsin

Lake Geneva, Wisconsin

Rockford, Illinois

Dallas, Oregon

Dallas, Oregon

Dallas, Oregon

Dallas, Oregon

Dallas, Oregon

Section 5. "Flat," "Box," or "Shed" Dormer

Although the name isn't very flattering these dormers have clean lines. They provide headroom that is more unobstructed so they don't have to be built as high. The pitch of the roof is important to assure proper water drainage or run-off.

Lewisville, Texas

Indian Wells, Texas

Indian Wells, Texas

123

Christi, Texas

Aransas Pass, Texas

Palacios, Texas

Palacios, Texas

Fallets Island, Texas

Galveston Island, Texas

Galveston, Texas

Florida Keys

125

Key West, Florida

Key West, Florida

Key West, Florida

Key West, Florida

Key West, Florida

Savannah, Georgia

Savannah, Georgia

127

Savannah

Savannah, Georgia

Savannah, Georgia

Hilton Head, South Carolina

Beaufort, South Carolina

Low Country near Beaufort, South Carolina

Charleston, South Carolina

Charleston, South Carolina

Charleston, South Carolina

Charleston, South Carolina

Folly Beach, south Carolina

Charleston, South Carolina

Ashville, North Carolina

Ashville, North Carolina

Rarity Ridge near Oak Ridge, Tennessee

Winchester, Kentucky

Winchester, Kentucky

Winchester, Kentucky

Frankfort, Kentucky

Frankfort, Kentucky

Evansville, Indiana

Fort Branch, Indiana

Oaktown, Indiana

Bloomington, Illinois

Bloomington, Illinois

Beloit, Wisconsin

Walworth, Wisconsin

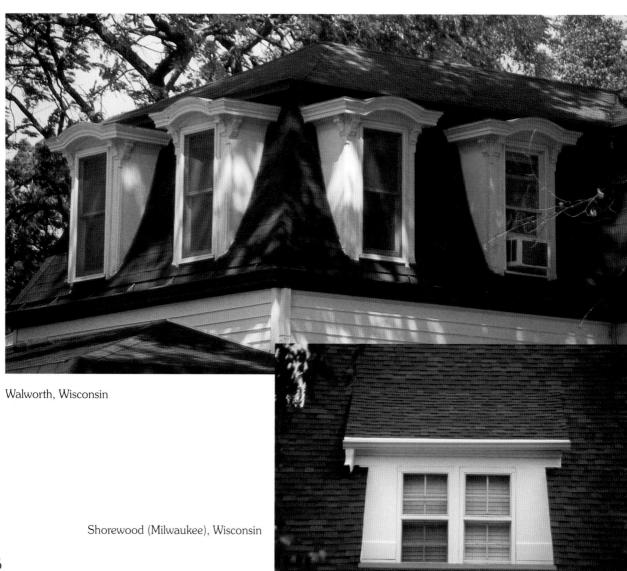

Walworth, Wisconsin

Shorewood (Milwaukee), Wisconsin

Monroe, Wisconsin

Rockford, Illinois

Wisconsin Dells, Wisconsin

Dallas, Oregon

Section 6. Turret Dormer

These dormers appear to be half mini-towers tucked into the roof. The rooflines and angles of the walls and windows can give the appearance of a castle to the exterior of the house.

Indian Wells, Texas

Indian Wells, Texas

Galveston Island, Texas

Lewisville, Texas. This dormer could also be called a small tower.

Galveston, Texas

Galveston, Texas

Key West, Florida

Hilton Head, South Carolina

Ashville, North Carolina

Ashville, North Carolina

Winchester, Kentucky

Winchester, Kentucky

Evansville, Indiana

Lexington, Kentucky

Bloomington, Illinois

Bloomington, Illinois

Beloit, Wisconsin

Bloomington, Illinois

Sharon, Wisconsin

Section 7. Deck Dormer

This dormer adds much diversification to a structure. Not only does it include all the regular benefits of a dormer, it provides access to the outdoors as well. It's a great dormer to use in areas where there is a view to be enjoyed. The rooflines on this style of dormer can be any one of those shown in previous sections.

Fallets Island, Texas. Deck dormer with hipped roof.

Fallets Island, Texas. Deck dormer with gable roof.

Fallets Island, Texas. Deck dormer with hipped roof.

Fallets Island, Texas. Deck dormer with "flat," "box," or "shed" roof.

Fallets Island, Texas. Through-the-cornice dormer with deck on top.

Fallets Island, Texas. Deck dormer with hipped roof.

Galveston Island, Texas. Deck dormer with turret roof.

Galveston Island, Texas. Deck dormer with "pediment" or "triangle."

Florida Keys. Deck dormer with gable roof.

Florida Keys. Deck dormer with gable roof.

Key West, Florida. Deck dormer with hipped roof.

Key West, Florida. Deck dormer with gable roof.

Key West, Florida. Deck dormer with gable roof.

Key West, Florida. Deck dormer with combined hip and gable roof.

Key West, Florida. Deck dormer with gable roof.

Key West, Florida. Deck dormer with gable roof.

Hilton Head, South Carolina. Deck dormer with "flat," "box," or "shed" roof.

Hilton Head, South Carolina. Deck dormer with "flat," "box," or "shed" roof.

Savannah, Georgia. Deck dormer with "pediment" or "triangle" gable roof.

Folly Beach, South Carolina. Deck dormer with enclosed eave gable roof.

Charleston, South Carolina. Deck dormer with gable roof.

Folly Beach, South Carolina. Deck dormer with hipped roof.

Flat Rock, North Carolina. Deck dormer with "flat," "box," or "shed" roof.

Ashville, North Carolina. Deck dormer with gable roof and awning.

Ashville, North Carolina. Deck dormer with combination pediment and "flat," "box," or "shed" roof.

Ashville, North Carolina. "Eyebrow" deck dormer.

Winchester, Kentucky. Deck dormer through-the-cornice with pediment gable roof.

Winchester, Kentucky. Deck dormer with hipped dormer.

Evansville, Indiana. Inset deck dormer.

Evansville, Indiana. Deck dormer with gable roof.

Bloomington, Illinois. Deck dormer with gable roof.

Bloomington, Illinois. Deck dormer pediment with gable roof.

Bloomington, Illinois. Deck dormer with hipped roof.

Bloomington, Illinois. Deck dormer with "flat," "box," or "shed" roof.

Bloomington, Illinois. Deck dormer within a "pediment" or "triangle" dormer.

South Beloit, Illinois. Deck dormer with enclosed eave gable roof.

Beloit, Wisconsin. Deck dormer with hipped roof.

Lake Geneva, Wisconsin. Deck dormer with gable roof.

Mitchell, South Dakota. Deck dormer with hipped roof.

Lake Geneva, Wisconsin. Deck dormer with hipped roof.

Dallas, Oregon. Deck dormer with "pediment" or "triangle" gable roof.

Section 8. Inset Dormer

Inset dormers were another rare find for me. The way they are tucked into the roof gives a cozy appearance to the structure. The rooflines on this style of dormer can be any one of those shown in previous sections.

Lewisville, Texas. Inset dormer with gable roof.

Indian Wells, Texas. Inset dormer with enclosed eave gable roof.

Indian Wells, Texas. Inset dormer with enclosed eave gable roof.

Fallets Island, Texas. Inset dormer with hipped roof.

Flat Rock, North Carolina. Inset dormer with gable roof.

155

Bloomington, Illinois. Inset dormer with gable roof.

Beloit, Wisconsin. Inset dormer with "pediment" or "triangle" roof.

Shorewood (Milwaukee), Wisconsin. Inset dormer with "flat", "box" or "shed" roof.

Dallas, Oregon. Inset dormer with "flat," "box," or "shed" roof.

Rockford, Illinois. Inset dormer with arched roof.

Dallas, Oregon. Inset dormer with gable roof.

Section 9. "Through-the-Cornice" or "Wall" Dormer

This style of dormer has a window that is flush with the structure's wall plane, and intersects the roof's cornice line. The dormer is more of a vertically projecting wall than an addition to the roof. The rooflines on this style of dormer can be any one of those shown in previous sections.

Lewisville, Texas. "Through-the-cornice" or "wall" dormer with enclosed eave gable roof.

Falcon, Texas. "Through-the-cornice" or "wall" dormer with gable roof.

Crawford, Texas. "Through-the-cornice" or "wall" dormer with arched roof.

Hildago, Texas. "Through-the-cornice" or "wall" dormer with enclosed eave gable roof.

Mission Texas. "Through-the-cornice" or "wall" dormer with enclosed eave gable roof.

Mission, Texas. "Through-the-cornice" or "wall" dormer with wrap-around enclosed eave gable roof.

Mission, Texas. "Through-the-cornice" or "wall" dormer with enclosed eave gable roof.

Mission, Texas. "Through-the-cornice" or "wall" dormer with enclosed eave gable roof.

Mission, Texas. "Through-the-cornice" or "wall" dormer with hipped roof.

Mission, Texas. "Through-the-cornice" or "wall" dormer with hipped roof.

Mission, Texas. "Through-the-cornice" or "wall" dormer with enclosed eave gable roof.

Mission, Texas. "Through-the-cornice" or "wall" dormer with enclosed eave gable roof.

Indian Wells, Texas. "Through-the-cornice" or "wall" dormer with enclosed eave gable roof.

Indian Wells, Texas. "Through-the-cornice" or "wall" dormer with enclosed eave gable roof.

Indian Wells, Texas. "Through-the-cornice" or "wall" dormer with enclosed eave gable roof.

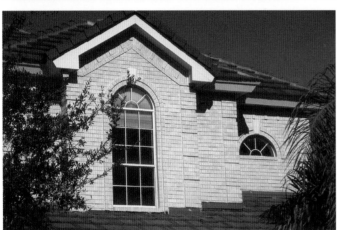

Indian Wells, Texas. "Through-the-cornice" or "wall" dormer with enclosed eave gable roof.

Indian Wells, Texas. "Through-the-cornice" or "wall" dormer with enclosed eave gable roof.

Indian Wells, Texas. "Through-the-cornice" or "wall" dormer with enclosed eave gable roof.

Indian Wells, Texas. "Through-the-cornice" or "wall" dormer with enclosed eave gable roof.

Brownsville, Texas. "Through-the-cornice" or "wall" dormer with arched roof.

Brownsville, Texas. "Through-the-cornice" or "wall" dormer with arched roof.

Brownsville, Texas. "Through-the-cornice" or "wall" dormer with arched roof.

Rockport, Texas. "Through-the-cornice" or "wall" dormer with enclosed eave gable roof.

Brownsville, Texas. "Through-the-cornice" or "wall" dormer with hipped roof.

Fallets Island, Texas. "Through-the-cornice" or "wall" dormer with gable roof.

Fallets Island, Texas. "Through-the-cornice" or "wall" dormer with flat roof.

Fallets Island, Texas. "Through-the-cornice" or "wall" dormer with flat roof.

Galveston Island, Texas. "Through-the-cornice" or "wall" dormer with hipped roof.

Galveston Island, Texas. "Through-the-cornice" or "wall" dormer with enclosed eave gable roof.

Galveston Island, Texas. "Through-the-cornice" or "wall" dormer with enclosed eave gable roof.

Galveston Island, Texas. "Through-the-cornice" or "wall" dormer with enclosed eave gable roof.

Galveston Island, Texas. "Through-the-cornice" or "wall" dormer with gable roof.

Florida Keys. "Through-the-cornice" or "wall" dormer with gable roof.

Galveston, Texas. "Through-the-cornice" or "wall" dormer with "pediment" or "triangle" gable roof.

Florida Keys. "Through-the-cornice" or "wall" dormer with gable roof.

Florida Keys. "Through-the-cornice" or "wall" dormer with gable roof.

Key West, Florida. "Through-the-cornice" or "wall" dormer with gable roof.

Florida Keys. "Through-the-cornice" or "wall" dormer with gable roof.

Florida Keys. "Through-the-cornice" or "wall" dormer with flat roof.

Florida Keys. "Through-the-cornice" or "wall" dormer with flat roof.

Savannah, Georgia. "Through-the-cornice" or "wall" dormer with modified gable roof.

Bluffton, South Carolina. "Through-the-cornice" or "wall" dormer with gable roof.

Hilton Head, South Carolina. "Through-the-cornice" or "wall" dormer with flat roof.

Low Country near Beaufort, Texas. "Through-the-cornice" or "wall" dormer with hipped roof.

Folly Beach, South Carolina. "Through-the-cornice" or "wall" dormer with gable roof.

Folly Beach, South Carolina. "Through-the-cornice" or "wall" dormer with gable roof.

Folly beach, South Carolina. "Through-the-cornice" or "wall" dormer with gable roof.

Folly Beach, South Carolina. "Through-the-cornice" or "wall" dormer with hipped roof.

Charleston, South Carolina. "Through-the-cornice" or "wall" dormer with gable roof.

Ashville, North Carolina. "Through-the-cornice" or "wall" dormer with enclosed eave gable roof.

Ashville, North Carolina. "Through-the-cornice" or "wall" dormer with arched roof.

Ashville, North Carolina. "Through-the-cornice" or "wall" dormer with gable roof.

Flat Rock, North Carolina. "Through-the-cornice" or "wall" dormer with gable roof.

Ashville, North Carolina. "Through-the-cornice" or "wall" dormer with modified roof.

Ashville, North Carolina. "Through-the-cornice" or "wall" dormer with gable roof.

Ashville, North Carolina. "Through-the-cornice" or "wall" dormer with gable roof.

Clinton, Tennessee. "Through-the-cornice" or "wall" dormer with enclosed eave gable roof.

Winchester, Kentucky. "Through-the-cornice" or "wall" dormer with broken pediment roof.

Winchester, Kentucky. "Through-the-cornice" or "wall" dormer with gable roof.

Winchester, Kentucky. "Through-the-cornice" or "wall" dormer with "pediment" or "triangle" gable roof.

Frankfort, Kentucky. "Through-the-cornice" or "wall" dormer with gable roof.

Winchester, Kentucky. "Through-the-cornice" or "wall" dormer with "pediment" or "triangle" gable roof.

Lexington, Kentucky. "Through-the-cornice" or "wall" dormer with modified gable roof.

Lexington, Kentucky. "Through-the-cornice" or "wall" dormer with modified gable roof.

Frankfort, Kentucky. "Through-the-cornice" or "wall" dormer with "pediment" or "triangle" gable roof.

Lexington, Kentucky. "Through-the-cornice" or "wall" dormer with enclosed eave gable roof.

Lexington, Kentucky. "Through-the-cornice" or "wall" dormer with enclosed eave gable roof.

Georgetown, Kentucky. "Through-the-cornice" or "wall" dormer with wrap-around enclosed eave gable roof.

Georgetown, Kentucky. "Through-the-cornice" or "wall" double dormer with enclosed eave gable roof.

Georgetown, Kentucky. "Through-the-cornice" or "wall" dormer with enclosed eave gable roof.

Georgetown, Kentucky. "Through-the-cornice" or "wall" dormer with enclosed eave gable roof.

Georgetown, Kentucky. "Through-the-cornice" or "wall" dormer with enclosed eave gable roof.

Evansville, Indiana. "Through-the-cornice" or "wall" dormer with gable roof.

Evansville, Indiana. "Through-the-cornice" or "wall" dormer with gable roof.

Evansville, Indiana. "Through-the-cornice" or "wall" dormer with gable roof.

Evansville, Indiana. "Through-the-cornice" or "wall" dormer with wrap-around enclosed eave gable roof.

Princeton, Indiana. "Through-the-cornice" or "wall" dormer with flat roof gable roof.

Princeton, Indiana. "Through-the-cornice" or "wall" dormer with gable roof enclosed eave roof.

Turkey Run State Park, Indiana. "Through-the-cornice" or "wall" dormer with pediment roof.

Danville, Illinois. "Through-the-cornice" or "wall" dormer with gable roof.

Danville, Illinois. "Through-the-cornice" or "wall" dormer with enclosed eave roof.

Bloomington, Illinois. "Through-the-cornice" or "wall" dormer with gable roof.

Bloomington, Illinois. "Through-the-cornice" or "wall" dormer with open pediment roof.

Bloomington, Illinois. "Through-the-cornice" or "wall" dormer with gable roof.

Bloomington, Illinois. "Through-the-cornice" or "wall" dormer with arched roof.

Bloomington, Illinois. "Through-the-cornice" or "wall" dormer with arched roof.

Orfordville, Wisconsin. "Through-the-cornice" or "wall" dormer with hipped roof.

Orfordville, Wisconsin. "Through-the-cornice" or "wall" dormer with hipped roof.

Beloit, Wisconsin. "Through-the-cornice" or "wall" dormer with enclosed eave gable roof.

Beloit, Wisconsin. "Through-the-cornice" or "wall" dormer with hipped and gable roof.

Beloit, Wisconsin. "Through-the-cornice" or "wall" dormer with gable roof.

Lake Geneva, Wisconsin. "Through-the-cornice" or "wall" dormer with gable roof.

Lake Geneva, Wisconsin. "Through-the-cornice" or "wall" dormer with enclosed eave gable roof.

Shorewood (Milwaukee), Wisconsin. "Through-the-cornice" or "wall" dormer with gable roof.

Shorewood (Milwaukee), Wisconsin. "Through-the-cornice" or "wall" dormer with gable roof.

Shorewood (Milwaukee), Wisconsin. "Through-the-cornice" or "wall" dormer with flat roof.

Shorewood (Milwaukee), Wisconsin. "Through-the-cornice" or "wall" dormer with gable roof.

Lake Geneva, Wisconsin. "Through-the-cornice" or "wall" dormer with gable roof.

Lake Geneva, Wisconsin. "Through-the-cornice" or "wall" dormer with thatch roof style.

Lake Geneva, Wisconsin. "Through-the-cornice" or "wall" dormer with gable roof.

Lake Geneva, Wisconsin. "Through-the-cornice" or "wall" dormer with gable roof.

Lake Geneva, Wisconsin. "Through-the-cornice" or "wall" dormer with gable roof.

Rockford, Illinois. "Through-the-cornice" or "wall" dormer with double gable roof with wrap-around eaves.

Rockford, Illinois. "Through-the-cornice" or "wall" dormer with enclosed eave gable roof.

Rockford, Illinois. "Through-the-cornice" or "wall" dormer with hipped roof.

Rockford, Illinois. "Through-the-cornice" or "wall" dormer with hipped roof.

Rockford, Illinois. "Through-the-cornice" or "wall" dormer with hipped roof.

Beloit, Wisconsin. "Through-the-cornice" or "wall" dormer with flat roof.

Beloit, Wisconsin. "Through-the-cornice" or "wall" dormer with pediment roof.

Beloit, Wisconsin. "Through-the-cornice" or "wall" dormer with pediment roof.

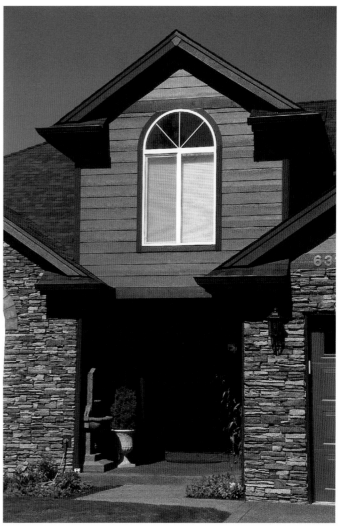

Dallas, Oregon. "Through-the-cornice" or "wall" dormer with wrap-around enclosed eave roof.

Dallas, Oregon. "Through-the-cornice" or "wall" dormer with gable roof.

Dallas, Oregon. "Through-the-cornice" or "wall" dormer with gable roof.

Section 10. Special Mention Dormers

The dormers in this section have roof lines or features that deserve special mention. They are unique in their design, some more so than others, and don't fall easily into any of the above categories. I've left them for last so readers will have had a chance to see standard dormer designs and will therefore better appreciate the unique aspects of these.

South Padre Island, Texas. This unusual dormer has no window or vent, but deserves special mention because of the roof line—a combination of hipped and gable dormer.

Bloomington, Illinois. The same holds true for the other dormer on the house, but it is even more unusual because of the chimney that goes through it.

Key West, Florida. The roof line on the dormer is gable. The front of the dormer has a Spanish influence.

Savannah, Georgia. Here is a dormer with a gable and flat top roof combination. Much special attention has been placed on the details.

Savannah, Georgia. This dormer has such a flair that it speaks for itself.

Beaufort, South Carolina. This dormer was given special treatment, with two gable roofs side by side, enclosed eaves, and arched windows.

Clinton, Tennessee. This little hipped dormer has been tucked in between two roofs.

185

Winchester, Kentucky. Here is a grand old turreted dormer that looks like a small tower.

Lexington, Kentucky. A combination pediment and arched through-the-cornice dormer that is distinctive in its design.

Winchester, Kentucky. This dormer design includes two side by side pedimented gable roofs, enclosed eaves, and two styles of windows.

Evansville, Indiana. Baroque inset dormers grace this ostentatious mansard roof.

Evansville, Indiana. A chimney penetrates and almost dominates this hipped dormer.

Evansville, Indiana. The protruding roofline of this pedimented gable dormer thrusts forward, providing cover for the deck below.

Evansville, Indiana. The protruding wrap-around enclosed eaves and two shapes of windows add flair to this dormer.

Evansville, Indiana. A front view of the pedimented gable dormer.

Evansville, Indiana. An exotic dormer peeks through the branches, allowing us a glimpse of its arched and flat roof and windows.

187

Fort Branch, Indiana. The combination of a hipped roof at the rear and a gable roof in front makes this dormer unique.

Bloomington, Illinois. A creative hipped dormer at the edge of the roof line.

Bloomington, Illinois. A creative dormer that bestows a cap for the peak of the roof.

Bloomington, Illinois. A pediment dormer and inset dormer combined.

Bloomington, Illinois. A through-the-cornice pediment dormer with a flamboyant dormer below.

Bloomington, Illinois. No windows or vents were used to decorate this unusually shaped dormer.

Oak Park (Chicago), Illinois. This dormer has the appearance of a bay window.

Bloomington, Illinois. Side by side gable dormers.

Oak Park (Chicago), Illinois. This pedimented gable dormer has rounded edges, round poles, and a recessed deck.

Oak Park (Chicago), Illinois. The finish along the sides creates a simple but ornate dormer.

Oak Park (Chicago), Illinois. A gable dormer with flat roof edges, stylish mixed window shapes, and creative detail.

Oak Park (Chicago), Illinois. This dormer has a combination hipped and gable roof, and is finished off with an arched window.

Beloit, Wisconsin. The detailed decoration on this gable dormer deserves special attention.

Shorewood (Milwaukee), Wisconsin. A flat roof dormer with a small gable roof and three paned glass windows trim this dormer.

Oak Park (Chicago), Illinois. A great combination of a small dormer at the peak of the roof, a deck dormer, and decorative colors make this an eye-catcher.

Lake Geneva, Wisconsin. The dormer is inset in the roof with a reverse arch at the base.

Monroe, Wisconsin. This dormer has a hipped roof and a through-the-cornice pediment. Those lines and the windows create a dormer that garnishes the roof of this house.

Beloit, Wisconsin. Three matching distinguished dormers grace this roof.

Rockford, Illinois. An especially slim gable dormer with a graceful window.

Beloit, Wisconsin. Close-up of one of the three matching dormers, to better view the detail.

Dallas, Oregon. A hipped dormer with a gable dormer on top.

Trempealeau, Wisconsin. The lines of the roof give these dormers an oriental flair.

Dallas, Oregon. This may be classified as a hanging or wall dormer, and is a grand way to add a vent.

Dallas, Oregon. A pediment dormer with a tower that matches the belvedere behind.

Independence, Oregon. A hipped dormer in front of a hipped dormer.